JADEDRA GILMORE-BARBER

THE
COUNTING
MERMAIDS

Published by: EarKanDee LLC
For more information: earkandee.educate@gmail.com
ISBN: 979-8-9925483-4-1
www.earkandeeonline.com

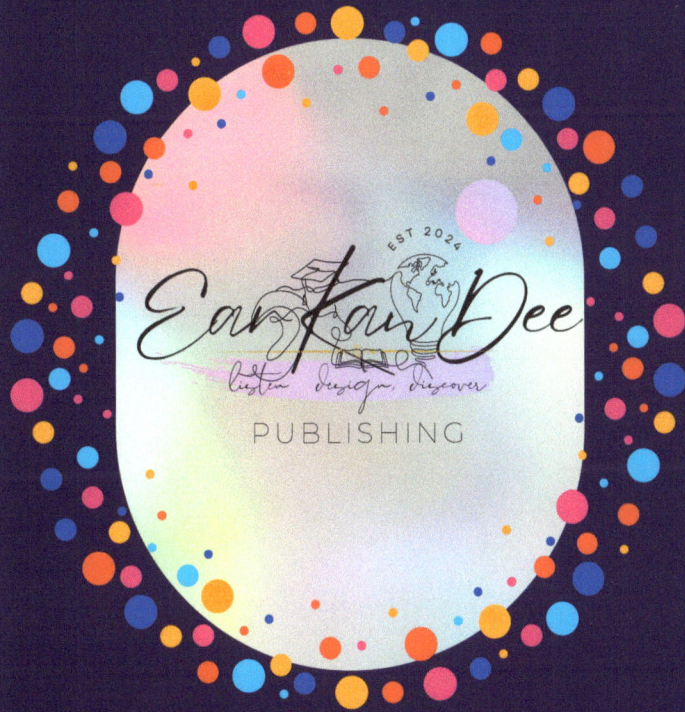

EarKanDee

listen, design, discover

PUBLISHING

EST 2024

"If better is possible, good is not enough."
— Benjamin Franklin

TABLE OF CONTENTS

NUMBERS 0-20

0 1 2 3 4 5 6 7 8 9

10 11 12 13 14 15 16 17 18

19 20

0
zero

Counting under the sea is always filled with fun!

I just saw a school of fish, but now there are zero. That means none!

2

1

one

One green
seahorse,
WOW, what
a sight!

I wonder where his family is
at this hour of the night.

2

two

Two tiny puffer fish going to join some friends

Hopefully they find them soon before the party ends!

4

3
three

<u>Three</u> baby
sea turtles
looking for
a treat

Found a group of sea sponges
and grabbed a bite to eat!

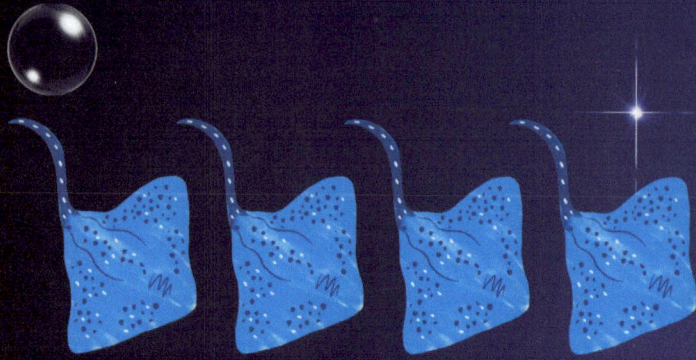

four

<u>Four</u> stellar

stingrays

lingering below

Looking for a place to sleep

where others rarely go.

6

5

five

Five wiggly jelly fish floating slowly by

I tried to say hello to them, but I guess they're very shy.

6

six

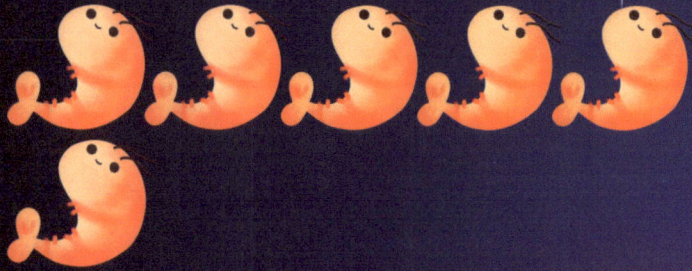

Six adorable silly shrimp playing copycat

Began to act like the mermaid by swimming on their back!

7
seven

Seven friendly dolphins splashing all around

Jumping up and doing tricks while making whistling sounds.

8

eight

Eight dazzling octopuses moving really fast

Racing is so fun, WATCH OUT! They might ink you if you pass!

10

9

nine

Nine snapping lobsters out enjoying the ocean view

Be careful not to get too close they may reach up and pinch you!

10 ten

Ten shining starfish happy as can be

They love to add joy to all that live beneath the sea.

12

11

eleven

Eleven fancy clams sitting gracefully in the sea

With gorgeous pearls for necklaces all mermaids love to see!

12
twelve

Twelve jolly clownfish joking all around

Made a mermaid laugh that was feeling rather down.

14

13

thirteen

Thirteen caring seals always out to do good deeds

Helped a mermaid friend when she shared with them her needs.

14
fourteen

Fourteen big and brave sharks set out to find a snack

Watch out for these hungry creatures, NEVER turn your back!

15
fifteen

Fifteen funny squid played a game of hide and seek

They closed their eyes, counted to 10, and did not take a peek!

16

sixteen

Sixteen prickly sea urchins feeling joyful and divine

Although they look quite similar, they're not a porcupine!

18

17
seventeen

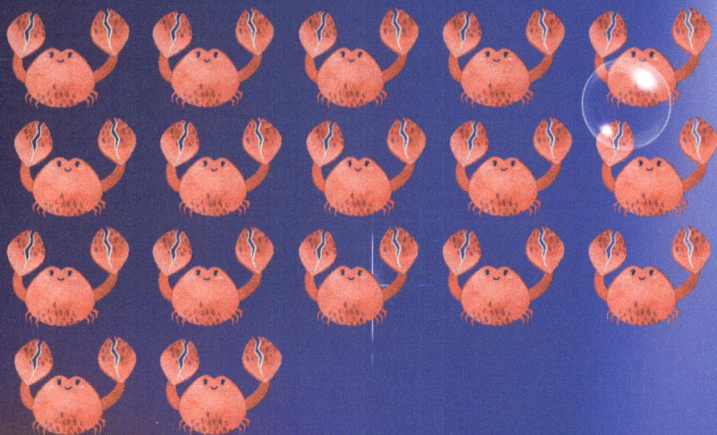

Seventeen red
crabs walk along
the ocean floor

The same way they move around
when traveling the shore.

18 **eighteen**

Eighteen peaceful orcas gliding through the sea

Roaming what seems to be an endless home so happy, wild, and free!

19
nineteen

Nineteen lovely lionfish, dainty and carefree

Know that true beauty comes from within your heart. That is a guarantee.

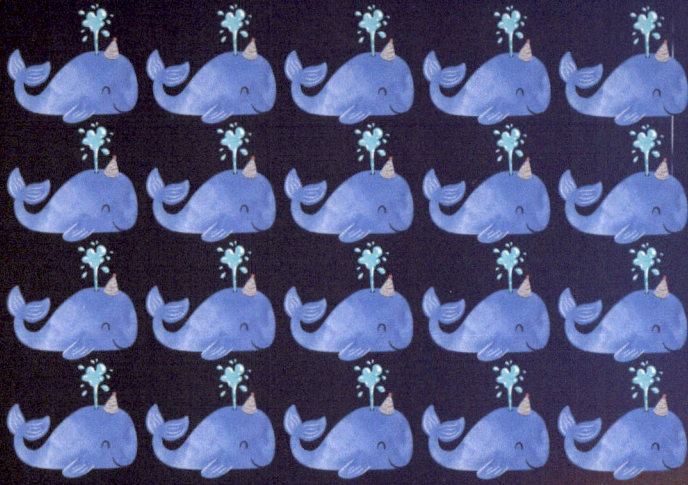

20

twenty

Twenty excited whales wearing hats with such delight

On their way to a party after receiving an invite!

Counting like a mermaid takes practice, not just luck.

So dive right in, embrace mistakes, and promise to never give up!

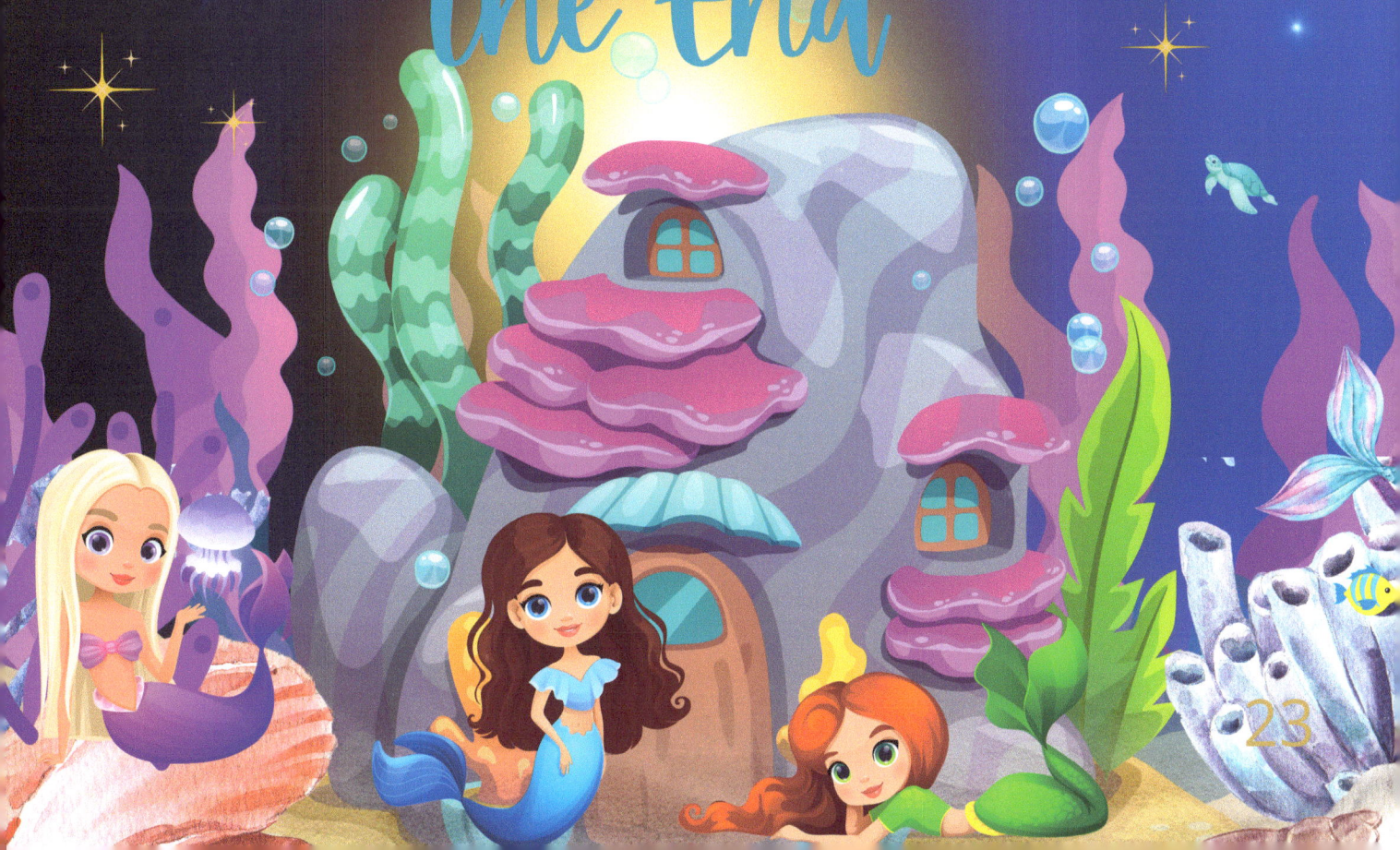

The End

Bonus Content: Coloring Pages

Biography

Dr. Jadedra Gilmore-Barber was born and raised in Fort Valley, Georgia. She has been an educator for over 14 years. Dr. Gilmore-Barber has held various roles during her years in the field of education. She has served as a teacher, academic coach, and administrator. She graduated from Georgia Southwestern State University with a bachelor's degree in Early Childhood Education, a master's degree in Curriculum and Instruction, and a specialist degree in Teacher Leadership. She graduated from Columbus State University with a specialist add-on in Educational Leadership, and a doctoral in Educational Leadership. Dr. Gilmore-Barber's philosophy of education is "Where better is possible, good is not enough!" She strives to motivate students to gain and maintain a love of learning.

EARKANDEE LLC

LISTEN | DESIGN | DISCOVER

"Where better is possible, good is not enough."

PUBLISHING "EFFECTIVE" EDUCATIONAL RESOURCES GLOBALLY

AVAILABLE
NOW ON AMAZON

"MUST LEARN SIGHTWORDS" THROUGH READING, WRITING, AND MATH WORKBOOK (KINDERGARTEN)
AVAILABLE NOW

"MUST LEARN SIGHTWORDS" THROUGH READING, WRITING, AND MATH WORKBOOK (FIRST GRADE)
AVAILABLE NOW

"MUST LEARN SIGHTWORDS" THROUGH READING, WRITING, AND MATH WORKBOOK (SECOND GRADE)
AVAILABLE NOW

MY FIRST COUNTING IN SPACE BOOK

AVAILABLE NOW

MATH "GROWTH" JOURNAL GRADE LEVELS K-5
AVAILABLE NOW

MATH "GROWTH" JOURNAL GRADE LEVELS 6-8
AVAILABLE NOW

MATH "GROWTH" JOURNAL GRADE LEVELS 9-12
AVAILABLE NOW

MY FIRST COUNTING IN SPORTS BOOK

AVAILABLE NOW

WRITING ON THE ROADS OF THE USA "STATES": K-2 HANDWRITING WORKBOOK
AVAILABLE NOW

G.R.E.A.T. IS FOR A DAY WITH MY GRANDPARENTS! CHILDREN'S BOOK
AVAILABLE NOW

THE MIGHTY MOVE UPSTAIRS CHILDREN'S BOOK
AVAILABLE NOW

MY FIRST COUNTING WITH MERMAIDS BOOK

AVAILABLE NOW

MY FIRST COUNTING WITH JUNGLE ANIMALS BOOK

AVAILABLE NOW

Stay Tuned For More....

DON'T FORGET TO REVIEW THIS WORKBOOK ON AMAZON

Find us at:
www.earkandeeonline.com

Contact us:
earkandee.educate@gmail.com

MY FIRST COUNTING IN THE SNOW BOOK

AVAILABLE NOW

www.ingramcontent.com/pod-product-compliance
Lightning Source LLC
Chambersburg PA
CBHW040232070426
42447CB00030B/158